I0060788

The Mechanisms of Evolution

A Critique of the Neo-Darwinian Modern Synthesis

Steven Hawkins

Light Press

Copyright © 2011 by Steven Hawkins

All rights reserved. This book, or parts thereof, may not be reproduced in any form without permission.

A catalogue record for this book is available from the British Library

ISBN: 978-1-907962-15-8

Published by Light Press

Reading, England

the interpretations surrounding the brute fact of evolution remain contentious, controversial, fractious, and acrimonious.[1]

Simon Conway Morris

Contents

Preface 7

Introduction 9

1   The Neo-Darwinian Modern Synthesis 13

2   Symbiogenesis 28

3   Developmental Systems Theory 45

4   The state of the Modern Synthesis 57

5   Do we understand the nature of the universe? 65

Bibliography 72

Preface

Most people alive today believe that the human species evolved from non-human planetary life-forms. Most people alive today also believe that the universe is an evolving entity. There are a number of very good reasons to believe both of these things. However, a very significant proportion of these people also believe something which is *not* supported by a plethora of good reasons. These people believe that they as individuals, and we as a species, have a very good understanding of how the evolutionary process works; that is to say, they believe that the sole (or the most important) mechanism of evolution is 'natural selection'. Today, people with such beliefs are typically advocates of the 'Neo-Darwinian Modern Synthesis'.

It is possible to believe that the sole (or the most important) mechanism of evolution is 'natural selection' whilst rejecting the 'Neo-Darwinian Modern Synthesis'. It is also possible to reject the idea that 'natural selection' is the sole (or the most important) mechanism of evolution and to propose an alternative dominant mechanism. Of course, it is also possible to believe in evolution whilst admitting that one is utterly clueless as to which mechanism is the sole (or the most important) mechanism of evolution.

This book has three objectives. The first objective is to consider a number of different possible mechanisms of evolution. The second objective is to consider which of these mechanisms, if any, plays the key role in the creation of new species. The third objective is to consider the possibility that even if we knew which mechanism, or mechanisms, underpins the evolutionary process that we still would not properly understand the nature of this process; this is because we do not understand the nature of the universe.

Introduction

The theory of evolution by common descent is now generally accepted as a brute fact. It was Charles Darwin who established beyond any reasonable doubt that evolution occurs. However, there are three meanings of evolution: *evolution as fact* – species are not fixed but arise out of and develop into other species, *evolution as path* – the actual routes that evolution has taken, and *evolution as mechanism* – the power that lies behind evolutionary change. Darwin himself had very little to say about the path of evolution, whilst the potency of his postulated evolutionary mechanism – 'natural selection' – has been seriously questioned.

In this book I will take it for granted that evolution is a fact. My objective is to critique various interpretations of 'evolution as path' and 'evolution as mechanism'. The current dominant paradigm of evolution is the 'modern synthesis': a synthesis of Darwinian natural selection with Mendelian genetic inheritance. This paradigm stresses that the *path* of evolution is gradualistic: very small cumulative changes in organisms give rise to adaptation. It also postulates that the *mechanism* of evolution is solely the natural selection of random gene mutations: all of the variation that gives rise to speciation and adaptation is argued to arise from the random mutation and reproductive shuffling of genes.

Various other positions have been suggested with regards to the path and the mechanisms of evolution. This is not surprising given that the fossil record itself gives no support to a gradualistic path, and that there have been no recorded cases of speciation through random mutation. I will be considering the arguments of two of the major contemporary alternative positions to see whether they make a convincing case against the 'modern synthesis' assumptions regarding 'evolution as path' and 'evolution as mechanism'. The first of the alternative positions is Developmental Systems Theory, which postulates that the object of natural selection is the whole life cycle of an organism. The second is Symbiogenesis, which postulates that natural selection has only a minor role to play in evolution.

In *Chapter One* I outline the claims of the dominant 'modern synthesis' paradigm, with the concentration being on 'selfish gene neo-Darwinism'. In *Chapter Two* and *Chapter Three* I outline the alternative positions, and I elucidate their arguments as to why the 'modern synthesis' is flawed in its claims regarding the path and/or the mechanism of evolution. Within these chapters I draw some preliminary conclusions as to whether the postulated alternative positions are justified, or alternatively, whether the 'modern synthesis' holds firm; it is possible that the alternative positions are just another way of looking at the same phenomena. In *Chapter Four* I draw together the various critiques so that conclusions can be made about the validity of the 'modern synthesis' claims regarding 'evolution as path' and 'evolution as mechanism'. In *Chapter Five* I consider the issue of whether, even if we did know which mechanism/mechanisms underpin the evolutionary process, that we still would not properly understand the nature of this process.

Chapter 1

The Neo-Darwinian Modern Synthesis

Charles Darwin proposed that the mechanism of evolution was natural variation and selection. He realized that more organisms are born than can survive and reproduce, and was thus led to hypothesize a struggle for survival. As there is obviously variation between organisms, those with advantageous traits will be those that survive. The heritable nature of the traits of the survivors leads to evolution, adaptation, and speciation. However, Darwin "could say little about the nature and causes of hereditary variation."[2]

It was the rediscovery of Gregor Mendel's model of genetic inheritance, in the early twentieth century, that provided a plausible model of hereditary variation. Mendel's "first law" asserts that the two alleles of each hereditary unit (gene) separate during gamete formation. His "second law" asserts that allele segregation in differing parts is wholly independent. It follows that there will be an immense amount of heritable variation in the gametes. The realization that this gene variation could be the basis for Darwinian natural selection led to the formulation of the neo-Darwinian 'modern synthesis'.

In the 'modern synthesis' natural selection is envisioned as acting on variation in persisting elements only. Populations, individual organisms and chromosomes do not persist; what does persist is small amounts of genetic material that are not broken up in meiosis. Richard Dawkins calls these stretches of genetic material the "immortal replicators". This focus on the selection of persisting elements means that there is a fundamental distinction in the 'modern synthesis' between the genotype/replicator on the one hand, and the phenotype/vehicle on the other.

Evolution as mechanism is envisioned as simply a change in the genetic composition of populations. Heredity occurs through transmission of chromosomal germ-line genes, which carry trait information. Variation results from allele shuffling in meiosis and random mutations, with each allele having a small phenotypic effect. Whilst selection of adapted phenotypes causes the increasing prevalence of their respective genotypes.

Within the 'modern synthesis' Dawkins has been a strong advocate of the view that the units of adaptive selection are genes rather than organisms. His theory of the "selfish gene" implies that sometimes the interests of the genes within an organism will be different to the interests of the organism itself. In *The Selfish Gene* Dawkins centers his arguments on the fragmenting effects of meiosis, which he argues is clear evidence that organisms cannot be replicators that natural selection works on. Natural selection requires persisting elements over many generations, organisms only last for a single generation. Dawkins stresses that genes have only *potential* immortality; a gene can last for millions of years or only a single generation. The reasons lying behind these differential survival rates of genes are central to his arguments:

> *Like successful Chicago gangsters, our genes have survived, in some cases for millions of years, in a highly competitive world. This entitles us to expect certain qualities in our genes. I shall argue that a predominant quality to be expected in a successful gene is ruthless selfishness. This gene selfishness will usually give rise to selfishness in individual behaviour.[3]*

> *The few new ones that succeed do so partly because they are lucky, but mainly because they have what it takes, and that means they are good at making survival machines. They have an effect on the embryonic development of each*

*successive body in which they find themselves,
such that that body is a little bit more likely to
live and reproduce than it would have been under the influence of the rival gene or allele.⁴*

So competition is between genes which are
vying for a chromosomal slot at the expense of their
alleles. Dawkins uses what he describes as a "fading
out" definition of a gene. The word "gene" refers to a
stretch of genetic material that survives in the form
of lots of copies for a significant period of evolutionary time. The smaller the stretch of genetic material,
the less likely it is to be divided, and the more copies
of it there are likely to be. This causes Dawkins to
assert that *The Selfish Gene* should be called: *"The
slightly selfish big bit of chromosome and the even
more selfish little bit of chromosome."*⁵

This gene's-eye view of allele competition and
multiple gene copies has two obvious implications.
Firstly, there is a selfish explanation for apparently
altruistic behaviour at the level of the organism:
"relatives share a substantial proportion of their
genes. Each selfish gene therefore has its loyalties
divided between different bodies."⁶ Secondly, there
will be Machiavellian *within* organism strategies as
genes via for survival: "There are even genes – called
mutators – that manipulate the rates of copying-
errors in other genes."⁷ The mutator genes selfishly

spread through the gene pool at the expense of those genes which are disadvantaged by the miscopying.

The question of the origin of heritable variation for the mechanism of selfish gene evolution needs to be highlighted. Dawkins claims that: "A body is the genes' way of preserving the genes unaltered."[8] This means that: "A monkey is a machine which preserves genes up trees, a fish is a machine which preserves genes in the water; there is even a small worm which preserves genes in German beer mats."[9] So evolution is an unwanted accident. The genes do not want to evolve new vehicles; they simply seek to preserve themselves. It is undesirable random mutations that have led to variation, adaptation, speciation, and evolution.

In *The Extended Phenotype* Dawkins expands and enriches the selfish-gene theory. He realizes that his concentration on meiosis in *The Selfish Gene* was missing the point because it implies that an asexually reproducing organism would be a replicator in itself, a giant gene. This means that:

> *Superficially, successive generations of stick-insect bodies appear to constitute a lineage of replicas. But if you change one member of the lineage (for instance by removing a leg), the change is not passed on down the lineage. By contrast, if you experimentally change one*

member of the lineage of genomes (for instance by X-rays), the change will be passed on down the lineage. This, rather than the fragmenting effects of meiosis, is the fundamental reason for saying that the individual organism is not the 'unit of selection' – not a true replicator.[10]

So the focus on genes as the units of selection in the 'modern synthesis' evolutionary mechanism rests on the Central Dogma of molecular biology – that information cannot move from proteins to DNA and RNA. Dawkins seeks to further justify his gene selectionism by arguing that this can best explain extended phenotypic effects. Prior to arguing for such effects Dawkins responds to those who have accused him of being a genetic determinist. He does this by stressing the difference between evolution and development. Evolution is an inflexible process of genes replicating themselves. Development concerns the influence of genes on phenotypes, which is exceedingly flexible. Thus:

A gene 'for' A in environment X may well turn out to be a gene for B in environment Y. It is simply meaningless to speak of an absolute, context-free, phenotypic effect of a given gene.[11]

Therefore, it is clear that Dawkins accepts that there is not a perfect correspondence between genotype and phenotype. However, he also claims that:

> *The statement, 'genes for performing behaviour X are favoured over genes for not performing X' has a vaguely naive and unprofessional ring to it...To say 'individuals that perform X are fitter than individuals that do not perform X' sounds much more respectable...But the two sentences are exactly equivalent in meaning. The second one says nothing that the first does not say more clearly.[12]*

So Dawkins claims that both the phenotypic effects of genes are environment-dependent, and that a phenotypic trait for X is 'exactly equivalent' to a gene for X. It is hard to see how a gene, if its phenotypic effect is environment-dependent, can in any meaningful way be said to be 'exactly equivalent' to a phenotypic trait. We will see that this point is the focus of the Developmental Systems Theorists belief that the 'modern synthesis' is preformationist. Dawkins attempts to make the statements consistent by arguing that we need to focus on *differences* and *relative* phenotypic effects. He argues that (with G = genotype, and P = phenotype):

there is a statistical tendency for individuals with G_1 to be more likely then individuals with G_2 to show P_1 (rather than P_2). Of course there is no need to demand that P_1 should always be associated with G_1, nor that G_1 should always lead to P_1: in the real world outside logic textbooks, the simple concepts of 'necessary' and 'sufficient' must usually be replaced by statistical equivalents.[13]

Therefore, Dawkins's claim that there are 'genes for performing behaviour X' is only a statistical tendency; whilst the 'exact equivalence' can only relate to one particular individual organism with a particular environmental history; it is a highly contingent equivalence. So, whilst Dawkins's position is that there is a statistical tendency for alleles likely to produce certain phenotypic effects to increase in the gene pool by natural selection, this is clearly an environmentally contingent tendency. Dawkins argues that there are several ways in which the effects of genes can extend outside their vehicles, and thereby increase their statistical prevalence in the gene pool.

The first of these extended phenotypic effects is the construction of animal artefacts. These artefacts – such as the construction of caddis-fly houses – are tools through which genes enhance their chances of getting into the next generation. There can also be shared interests in a single artefact, as when a beaver dam is the extended phenotype of several beavers.

The second group of effects relate to parasites and their hosts. Dawkins claims that, "it is logically sensible to regard parasite genes as having phenotypic expression in host bodies and behaviour."[14] An example of this is the case of the 'brainworm' *Dicrocelium dendriticium*. The 'brainworm' burrows into the suboesophagael ganglion of an ant's brain and is thereby able to control the ant's behaviour. The result is that, "infected ants climb to the top of grass stems, clamp their jaws in the plant and remain immobile as if asleep."[15] This is maladaptive behaviour for the ant – the uninfected ants have retreated to their nest – but it enhances the worm's chances of being eaten by its definitive host.

The final group of extended phenotypic effects is 'action at a distance'. An example is the "Bruce Effect" which is an effect that a male mouse can have on a female mouse that has just been inseminated by another male; chemical exposure means that the pregnancy is blocked. This means that: "Abortion in female mice, according to this hypothesis, is a phenotypic effect of a gene in male mice."[16] A further example is the cuckoo which, "uses a supernormally bright gape to inject its control into the reed warbler's nervous system via the eyes. It uses an especially loud begging cry to control the reed warbler's nervous system via the ears. Cuckoo genes, in exerting their developmental power over host phenotypes, have to rely on action at a distance."[17]

The central claim of the extended phenotype theory is that: "An animal's behaviour tends to maximize the survival of the genes 'for' that behaviour, whether or not those genes happen to be in the body of the particular animal performing it."[18] The range of cases that Dawkins refers to gives strong support for this claim. In many cases the behaviour of an organism is maladaptive in terms of its survival chances, whilst being adaptive for the genes that initiate the behaviour. It follows that the unit of selection is perhaps better thought of as the gene rather than the organism.

So the mechanism of evolution in the neo-Darwinian 'modern synthesis' is the unwelcome mutation of genes, which can be thought of as selfish, and as having extended phenotypic effects. A crucial aspect of this mechanism, which relates 'evolution as mechanism' to 'evolution as path', is that evolution proceeds in a gradualistic manner through accumulating small beneficial changes over time. This causes Dawkins to argue that:

Each successive change in the gradual evolu-
tionary process was simple enough, relative to
its predecessor, to have arisen by chance. But
the whole sequence of cumulative steps consti-
tutes anything but a chance process, when you
consider the complexity of the final end-product
relative to the original starting point. The cu-
mulative process is directed by non-random
survival.[19]

Dawkins argues that the alternative is that a
single macromutation could turn bare skin into a
fully functioning human eye. This is as likely as "a
hurricane blowing though a junkyard and chancing
to assemble a Boeing 747."[20] This is 'saltationism'
and Dawkins has to two arguments against it.
Firstly, the larger a mutation is the more likely it is
to be harmful: "if we consider mutations of ever-
increasing magnitude, there will come a point when,
the larger the mutation is, the less likely it is to be
beneficial; while if we consider mutations of ever-
decreasing magnitude, there will come a point when
the chance of a mutation's being beneficial is 50 per
cent."[21] Secondly, the bare skin – eye macromutation entails such a "large number of improvements,
their joint occurrence becomes so improbable as to
be, to all intents and purposes, impossible."[22]

So in the 'modern synthesis' adaptive evolutionary change comes from very small favourable random gene mutations that are accumulated over time. Natural selection is a constructive force because: "genes will be favoured if they are good at cooperating with other genes in the same gene pool,"[23] and because, "arms races... [propel] evolution in directions that we recognize as 'progressive', complex 'design'."[24] Any large macromutations will be harmful and will thus be eliminated by negative selection. This evolutionary mechanism leads to expectations of 'evolution as path'.

The actual evolutionary paths that occurred in the past are clearly a matter of speculation. The 'modern synthesis' asserts that there was gradualistic change with small accumulated adaptive traits forming in response to geographical isolation, and after sufficient time leading to speciation. There was a common ancestor, and there has been common descent through purely branching phylogenies ever since. This implies that our initial expectations of the fossil record would be to find a smooth and gradual transition between species, assuming of course that the record was complete.

However, in actuality the fossil record reveals trends that are extremely jerky, far from the expected smoothness. It was assumed by Darwin that the actual fossil record is jerky because it is incomplete, and that a complete record would show a smooth transition between species. There is another alternative, because, as Dawkins claims: "It is conceivable that there really never were any intermediates; conceivable that large evolutionary changes took place in a single generation."[25] In the discussion on gradualism we saw that Dawkins finds this inconceivable because it relies on saltationist macromutations which he claims are certain to be unfavourable.

The theory of punctuated equilibrium – originated by Niles Eldredge and Stephen Jay Gould – has often been cited as evidence against gradualism. In the punctuated equilibrium theory evolution occurs in sudden bursts which are punctuated by long periods of stasis. These 'bursts' have often been associated with 'saltationism', but Dawkins argues that they are two very distinct things; 'bursts' in the fossil record could equate to thousands of generations of a creature's existence. Dawkins does acknowledge that Eldredge and Gould: "saw analogies between themselves and the old schools of 'catastrophism' and 'saltationism'."[26] However, because Dawkins personally finds this inconceivable he claims that: "Eldredge and Gould are not saltationists."[27] Despite their own analogies, they *must* have only been talking about differential speeds of evolutionary gradualism.

So, how does the 'modern synthesis' explain the fossil record? Dawkins argues that an incomplete fossil record is exactly what you would expect if gradualistic evolution by natural selection occurred. This is because in the 'modern synthesis' geographical separation, "is the main process by which new species come into existence."[28] It follows that the fossils at any one location will not reveal information about an 'evolutionary event', but about a 'migrational event'.

Dawkins asks us to imagine an ancestral species in one location, and then postulates that a few members of the species become geographically isolated; we can then derive expectations of what should be in the fossil record. The isolated members will experience a different environment and will therefore accumulate random mutations that the non-isolated members do not; after sufficient time the isolated members will become a new species. A change that enables the geographical isolation to be transcended will enable the isolated members (now a new species) to return to their ancestral home; they could outcompete the ancestral species and replace them. Clearly, the fossil record at this location would be expected to show a jerky transition from the ancestral species to the descendent species.[29]

So, to recapitulate, in the 'modern synthesis' the path of evolution is one of branching phylogenies, which largely result from geographical separation. Changing conditions and 'migrational events' mean that the fossil record will not be smooth. It seems that either a perfectly smooth or an extremely jerky fossil record would provide support for the 'modern synthesis'.

Chapter 2

Symbiogenesis

The theory of symbiogenesis postulates a radically different conception of both 'evolution as mechanism' and 'evolution as path' to that of the 'modern synthesis'. Despite this radically different view the nature of the difference is actually more of a quantitative than a qualitative one. The 'modern synthesis' accepts that symbiosis occurs, but asserts that it has very little relevance to heritable variation, speciation, and evolution; it is the natural selection of random mutations that is the driving force behind these phenomena. In contrast, the symbiogenesis theory asserts that it is natural selection that is the minor partner; symbiogenesis is the driving force behind speciation and evolution, it is the provider of Darwin's heritable variation.

It is important to be clear on the difference between symbioses and symbiogenesis. Symbioses are, "long-term physical associations. Different types of organisms stick together and fuse to make a third kind of organism."[30] These associations are very common and are accepted by the 'modern synthesis'. Symbiogenesis is something that can result from a symbiosis. As Lynn Margulis and Dorion Sagan express it:

As members of two species respond over time to each other's presence, exploitative relationships may eventually become convivial to the point where neither organism exists without the other. Long-term stable symbioses that leads to evolutionary change is called "symbiogenesis." These mergers, long-term biological fusions beginning as symbiosis, are the engine of species evolution.[31]

One of the forerunners of the view that symbiogenesis could be the engine of speciation and heritable variation, is, perhaps paradoxically, Richard Dawkins. In *The Selfish Gene* and *The Extended Phenotype* Dawkins sights many examples of symbiosis, and even claims that the symbiotic origin of cells; "is one of those revolutionary ideas which it takes time to get used to, but it is an idea whose time has come."[32] Perhaps Dawkins is right – the time for revolution has come, and the objective of the revolution is to replace the neo-Darwinian 'modern synthesis'. Dawkins claims that:

> *Symbiotic relationships of mutual benefit are common among animals and plants. A lichen appears superficially to be an individual plant like any other. But it is really an intimate symbiotic union between a fungus and a green algae. Neither partner could live without the other. If their union had become just a bit more intimate we would no longer have been able to tell that a lichen was a double organism at all. Perhaps then there are other double or multiple organisms which we have not recognized as such. Perhaps even we ourselves?[33]*

This is a good statement of how symbiosis can lead to symbiogenesis, how the intimate union of different species gives rise to a new species. Symbiogenesis is the 'becoming just a bit more intimate'. We will see that the Symbiogenesis Theory claims that not just "we ourselves", but that *every* species on the planet is a symbiotic union. Dawkins himself goes even further than the symbiogenesists, who claim that the eukaryotic cell is symbiotic union. For Dawkins, "We are gigantic colonies of symbiotic genes."[34] It is quite strange that Dawkins can claim that genes are symbiotic, cells are symbiotic, that we ourselves are perhaps symbiotic organisms, and then not conclude that symbiosis is the dominant force behind variation and speciation. He even realizes that two organisms can fuse into one. In *The*

Extended Phenotype he approvingly cites Smith, who claims that:

> *In the cell habitat, an invading organism can progressively lose pieces of itself, slowly blending into the general background, its former existence betrayed only by some relic. Indeed, one is reminded of Alice in Wonderland's encounter with the Cheshire Cat. As she watched it, "it vanished quite slowly, beginning with the tail, and ending with the grin, which remained sometime after the rest of it had gone."[35]*

This passage describes full-blown symbiogenesis – the origin of a new species. It is time to look at the exact claims of the symbiogenesis theory from the perspectives of 'evolution as mechanism' and 'evolution as path'. Earlier in the book we noted that Darwin had little to say about the causes of heritable variation, and that the 'modern synthesis' postulates that this variation comes from random gene mutations. These mutations when they are correlated with geographical separation give rise to new species. In contrast, symbiogenesis theory claims that: "random mutation is wildly overemphasized as a source of hereditary variation"[36], and that "most evolutionary novelty arose and still arises directly from symbiosis."[37] Margulis and Sagan expand these claims in the following passages:

Mutations, genetic changes in live organisms, are inducible; this can be done by X-ray radiation or by addition of mutagenic chemicals to food. Many ways to induce mutations are known but none lead to new organisms. Mutation accumulation does not lead to new species or even to new organs or new tissues. If the egg and a batch of sperm of a mammal is subjected to mutation, yes, hereditary changes occur, but...99.9 per cent of the mutations are deleterious.[38]

the major source of inherited variation is not random mutation. Rather, the important transmitted variation that leads to evolutionary novelty comes from the acquisition of genomes. Entire sets of genes, indeed whole organisms each with its own genome, are acquired and incorporated by others. The most common route of genome acquisition, furthermore, is by the process known as symbiogenesis.[39]

So symbiogenesists claim that it is the acquisition of genomes, rather than random mutation that leads to speciation. The central mechanism of the 'modern synthesis' gets demoted: "Although random mutations influenced the course of evolution, their influence was mainly by loss, alteration, and refinement."[40] It is useful to go back to the origins of life itself in order to appreciate this fundamental difference between symbiogenesis and the 'modern synthesis'.

Dawkins postulates that replicators came first and that the cell originated in their service: "replicators perhaps discovered how to protect themselves, either chemically, or by building a physical wall of protein around themselves. This may have been how the first living cells appeared."[41] The cells are survival machines which "began as passive receptacles for the genes, providing little more than walls to protect them from the chemical warfare of their rivals and the ravages of accidental molecular bombardment."[42] Clearly, this view of the origins of life – that the cell is protective armour to help win a war – is going to lead to the construction of a selfish story.

In contrast, Margulis postulates that the cell membrane originated *first*, and that the replicators came *second*: "cell-like membranous enclosures form as naturally as bubbles when oil is shaken with water...Prelife, with a suitable source of energy inside a greasy membrane, grew chemically complex...After a great deal of metabolic evolution, which I believe occurred inside the self-maintaining greasy membrane, some, those containing phosphate and nucleosides with phosphate attached to them, acquired the ability to replicate more or less accurately."[43] In this account, there is no need for protective armour; the account doesn't establish a competition-cooperation dichotomy.

It is claimed by Margulis and Sagan that these first bacterial cells whilst they marked the origins of life, did not mark the origin of species. For them genetic material is so easily transferred between *all* the bacteria in the world that it is nonsensical to say that there are species of bacteria; speciation is solely a property of nucleated organisms. They are thus able to claim that: "The creative force of symbiosis produced eukaryotic cells from bacteria. Hence all larger organisms – protoctists, fungi, animals, and plants – originated symbiotically."[44] So they claim that as the eukaryotic cell itself is a symbiotic construction (which Dawkins claimed was an idea whose time had come), and that eukaryotic cells are definitive of a species, this means that all species are symbiotic constructions.

This claim by itself is perhaps not that radical – a giraffe is a 'symbiotic' creation because it is composed of symbiotically created cells. However, eukaryotic cell formation is just 'stage one' of the symbiogenetic theory. 'Stage two' concerns all of the evolution of life since this time: "Details abound that support the concept that all visible organisms, plants, animals, and fungi evolved by "body fusion". Fusion at the microscopic level led to genetic integration and formation of ever-more complex individuals."[45]

The focus of the symbiogenetic evolutionary mechanism is the microcosm. Microbes – which have uniquely capable complete genomes – drive evolutionary change forward. These genomes "come neatly packaged with long histories of heritable virtuosities and synthetic tricks. They provide just what is needed for an organism to change drastically and yet remain coherent and viable."[46] Initially, there is an association between two genomes, this can lead to a partnership, a symbiosis, and finally to a new species formed by symbiogenesis. While the 'modern synthesis' mutations are random, this process is driven by the interactions of organisms:

For two different types of genomes to merge and form a new one, the organisms themselves must have a reason to come together. Reasons vary. Organism A may find B delicious, and try to swallow B. Alternatively, organism A may require the chemical form of nitrogen excreted in the waste of B. Organism A may simply bask, at first, in the shade provided by B – or B may sequester the alkaline moisture that exudes at dawn from the pores of B. These are ecological issues with many subtleties, but they underlie the transfer and eventual merger of microbial genomes to larger forms of life.[47]

The result of this evolutionary mechanism is a new and comprehensive definition of a species: "Two live beings belong to the same species when the content and the number of integrated, formerly independent genomes that constitute them are the same."[48] There is a four level hierarchy of 'symbiotic partner integration' that leads to symbiogenetic speciation. The four levels are behavioural integration, metabolic integration, gene-product association, and genetic integration.

The behavioural integration of two symbiotic partners is the most basic and superficial type of symbiotic integration. Ivan Emmanuel Wallin in *Symbionticism and the Origin of Species* has named this level of integration "prototaxis" – the innate tendency of one kind of cell or organism to respond in a specific manner to another sort of organism. So two partners find themselves in the same place at the same time, and their behaviour towards each other is determined by ancestry and contingency. An example is the origin of plants:

No alga or plant ever evolved photosynthesis on it own. All shared some ancestor – recent or remote – that ate but failed to digest a green or red or greenish blue bacterial photosynthesizer. Prototaxis, in this case, is the tendency toward hunger on the part of the eater and toward resistance to digestion on the part of the eaten. Starvation in the light and resistance to digestion, in short, have led over and again to permanently pigmented photosynthetic organisms: Algae, lichens, plants, green worms, green hydra, brown corals, and giant clams.[49]

The *metabolic* level is the next most intimate level of symbiotic partner integration. Typically, the metabolic product of one partner is the food of the other; there is a two-way exchange of these products which over time leads to dependency and a loss of individuality. Examples are 'green animals' and lichens. One species of 'green animal' is the flatworm *Convoluta roscoffersis* which is a symbiotic integration of a worm with photosynthetic microorganisms. The photosynthetic products of the microorganisms become food for the worm, whilst the worm produces nitrogen-rich waste which is required by the microorganisms.

The third level of partner integration is "gene-product association". In this case proteins or RNA molecules from one partner are required for the functioning of the other. A good example is the symbiotic integration between a pea plant and its nitrogen-fixing bacterial nodules. In the partnership the production of hemoglobin is chimeric – the heme is produced by the bacteria and the globin is produced by the plant. So the hemoglobin which is required for the continued survival of both 'partners' is a result of their gene-product integration. The highest level of symbiotic intimacy is genetic integration. This occurs when a gene of one organism enters and remains with the genes of another, for example, when the gene of a free-living bacterium gets integrated into a plant nucleus.

There will tend to be a movement from the lowest level of integration toward the highest if this is in the interests of both partners. The 'interests' will be determined by natural selection. In some cases behavioural integration will not be transcended by deeper levels of integration, because this level maximises the offspring of the partners. It follows that: "The complete, irreversible integration of two different beings to form a new one will occur if at all times the physically associated organisms leave more descendents that do the independent unassociated ones."[50] The link between symbiogenesis and natural selection is made increasingly clear by the following passage:

The hundreds of mitochondria...in each of our cells, never leave these cells. Why? Because the world of animal tissue if full of oxygen and requires a flow of oxygen to the cell at all times. In the oxygen-rich Proterozoic, the cells that retained their mitochondria throughout their lives must have been "naturally selected" over those that occasionally let their mitochondria return to the bacterial world.[51]

The relationship between symbioses, symbiogenesis, random mutations, and natural selection, should now be clear: *"Protracted symbioses lead to symbiogenesis*: the origin of new organelles, organellar systems, tissues, organs, organisms, and species...*Symbiogenesis*, the inheritance of acquired genomes, mostly those of bacteria and other microbes, is the greatest source of evolutionary innovation...*Random mutations* only refine and alter, but do not produce, species-level change...*Natural selection* directs evolution through propagation and elimination of what it has already."[52]

So, in the words of Donald I. Williamson, there is "a saltational process in animal evolution which operates independently of the accumulation of mutations and then selection."[53] Williamson's work on invertebrates led him to conclude that animal-animal genome transfer is the only thing that can explain the distribution of larval types, as the genome that determines the animal larval form is different form that which determines the adult form. In other words, there have been numerous successful sexual encounters between members of different lineages: "Such successful matings between very distantly related animals occurred infrequently, some thirty to fifty times in 541 million years. This means a fertile, successful outcome happens roughly once in 10 million years."[54]

These encounters occurred by either external or internal fertilization and can, Williamson claims, explain why extremely different adults develop from nearly identical larvae, whereas closely related adults can develop from entirely different larvae. He concludes that: "all species that produce larvae, even caterpillars and other terrestrial animals, acquired foreign genomes at some point in their history."[55] So the sergestid shrimp is evolutionarily successful largely because it, "acquired, integrated, and put to work at least four intact genomes."[56] Random mutations have only a very minor explanatory role in the success of the shrimp.

So, evolution as mechanism in the symbiogenesis theory operates through the acquisition of genomes. When it comes to the higher taxa Margulis and Sagan admit that: "The inference of speciation and the evolution of higher taxa by symbiogenesis is often fraught with difficulty."[57] They suggest that mammalian speciation may have occurred through the Todd-Kolwicki Karyotypic Fission Theory. Under this theory the karyotype details – the total number and the morphology of the entire chromosome set of an organism – that have been collected clearly indicate saltationist speciation. This is because: "If Darwinian gradualism explains the origins of animal and plant species, it follows that closely related species should have similar karyotypes. They don't."[58]

The case for the symbiogenetic speciation of mammals is clearly less strong than for the lower taxa. However, it needs to be remembered that the higher taxa only constitute a very small minority of the planet's life-forms. Furthermore, whilst there are no cases of speciation by random mutation, there are two cases of speciation by symbiogenesis. The first of these is Theodosius Dobzhansky's experiments with *Drosophila*. Dobzhansky bred one group of fruit flies for two years at progressively hotter temperatures, at which point they could not produce fertile offspring with the group that had been breeding at normal temperatures. This was because symbiogenetic speciation had occurred: "the hot-breeding flies lacked an intracellular symbiotic bacteria found in the cold breeders."[59]

The second case of symbiotic speciation is Kwang Jeon's experiments with amoebae. Following an accident where some new amoebae infected his amoebae collection with a rod-shaped bacterium most of his collection died. However, a few survived with the bacteria inside them. These amoeba, "were easily killed by antibiotics, which, while deadly to bacteria, did not harm his normal "nonbacterized" amoebae. A change was occurring. The two types of organisms, bacteria and amoebae, were becoming one."[60] By switching nuclei between normal and infected cells Jeon was able to show that the infected nuclei needed a bacteria infected cell in order to survive.

It is time to turn to 'evolution as path'; it is clear that symbiogenesis involves a very different view to the 'modern synthesis'. The two main areas of difference are the shape of the phylogenetic evolutionary tree, and the presence of 'saltationism' in the evolutionary tree. Whilst the 'modern synthesis' envisions the phylogenetic tree as forever branching from a common ancestor, for the symbiogenesist: "the acquisition of heritable genomes can be depicted as an anastomosis, a fusing of branches."[61] Margulis and Sagan describe this process of anastomosis quite graphically:

> *Animal evolution resembles the evolution of machines, where the typewriters and televisionlike screens integrate to form laptops, and internal combustion engines and carriages merge to form automobiles. The principle stays the same: Well-honed parts integrate into startling new wholes.*[62]

The second area of difference is 'saltationism'. Symbiogenesis is clearly a saltationist theory of evolution. Whilst Dawkins argues against saltationism on the basis of the overwhelmingly likely deleterious effects of a macromutation, these arguments clearly do not apply to saltationism initiated by genome acquisition. Not surprisingly, Margulis and Sagan appeal to the jerky nature of the fossil record as support for symbiogenetic saltationism: "Punctuated equilibrium is there for all who take the time to see it. The discontinuous record of past life shows clearly that the transition from one species to another occurs in discrete jumps. In trilobites, snails, seed ferns, horses, lungfish, sharks, and clams, evidence abounds for punctuated change."[63]

We saw that Dawkins argues that punctuated equilibrium is just about *speeds* of evolution, not about saltationism. He also argues that the fossil record only reveals 'migrational events' and not 'evolutionary events'. In contrast, Margulis and Sagan cite Niles Eldredge's work with the well preserved Cambrian trilobites:

> *one species would continue with minor random variations for 800,000 years. Another would abruptly begin and overlie the first for 1.3 million years. The search for intermediate forms and gradual evolutionary change between the two species was always futile. The sedimentary rocks in which the glorious fossil record is embedded do not lie. They do not deceive. The record was punctuated, and the differences*

between species of extinct animals trapped in it were clean and distinct.[64]

There are clearly at least three interpretations that can be made. Firstly, the Darwinian line that the intermediates are missing. Secondly, the Dawksian refinement that the fossil record only records a 'migrational event'. Thirdly, that there were no intermediates and the fossil record reveals a symbiogenetic speciation. For now, we can at least conclude that the fossil record is concordant with the symbiogenesis paradigm, and its claim that evolution is punctuated, saltational, and driven by microbial mergers. Margulis and Sagan claim that: "The reliance on accumulation of random mutations in DNA is not so much "wrong" as oversimplified and incomplete: It misses the symbiotic forest for the genetic trees."[65] Similar sentiments on the importance of random mutations in evolution are expressed by the developmental systems theorists.

Chapter 3

Developmental Systems Theory

In the 'modern synthesis' the mechanism of evolution is natural selection acting on genes; genes are singled out because they are argued to be the only long-term persisting elements of organisms, and because they contain 'information' that leads to the development of favourable phenotypic traits. The symbiogenetic theory rejects this view of evolution because it denies that natural selection is the dominant force in generating speciation and heritable variation. In contrast, developmental systems theorists reject the 'modern synthesis' because they argue that natural selection acts on the whole life-cycle of an organism; it is meaningless to single out the genes as having more causal influence than any other factor that affects an organism.

By not giving priority to any factor in the developmental and evolutionary process developmental systems theorists claim to have dissolved several long-standing dichotomies: the innate/acquired distinction, internal versus external causation, and the nature/nurture opposition. They claim that the 'modern synthesis' is preformationist, as is any paradigm that accepts such dichotomies. Susan Oyama, Paul E. Griffiths and Russell D. Gray pose the question: "Can we shape our destiny, or are we robots programmed by our selfish genes?"[66] This question is framed in a slightly misleading manner; we have seen that 'selfish' genes do no imply that organisms are programmed robots devoid of free will. The real point they are trying to make is better expressed in the following passage:

> *The standard response to nature/nurture oppositions is the homily that nowadays everyone is an interactionist: All phenotypes are the joint product of genes and environment. According to one version of this conventional "interactionist" position, the real debate should not be about whether a particular trait is due to nature or nurture, but rather how much each "influences" the trait.*[67]

So, their critique of Dawksian selfish genes is not that it implies genetic determinism – as the first quote implies. The real claim is that the 'modern synthesis', whilst accepting that environmental factors and genetic factors interact to produce evolutionary outcomes, still sees the genes as containing information that will lead to a given outcome if the background environmental resources are available. This is even clearer in the following passage from Oyama: "Some might say, for instance, that the nature/nurture opposition is nonsensical because some things don't just mature, but require interaction. Here it is evident that *interaction*, far from challenging the concept of internally driven maturation, assumes it."[68]

So, Oyama claims that it is not appropriate to describe genes as "containing information" – information is the result of ontogeny rather than its cause. The whole developmental process is a life cycle that contains a diverse range of heritable resources and other non-inheritable resources, which are reconstructed in every cycle through "self-organization". It follows that the 'modern synthesis' distinction between replicators and vehicles is inappropriate. As Bruce H. Weber and David J. Depew argue:

> *DST champions infer from the presumptive causal parity of all developmental resources that the replicator/interactor distinction, on which the units of selection debate has been predicated, is ill conceived. When phenotypic traits are construed as developmental resources it can be seen that they are as much replicators as interactors.*[69]

A central part of the case for developmental systems theory is that a whole range of resources are heritable; the 'modern synthesis' with its sole focus on genes ignores these other resources and holds that they are not relevant to evolution. I will now outline this expanded range of heritable resources and explain why developmental systems theorists argue that they play an important role in evolution. Eva Jablonka claims that:

> *According to genic neo-Darwinism, nucleic acids are the sole units of heritable variation, the transmission of these units is independent of their expression, and the generation of genetic variations is not adaptively guided by the selective environment or the developmental history of the organism. This replicator-centered, gene-derived view of heredity is, however, not only severely limited but also severely misleading. There are multiple inheritance systems, with several modes of transmission for each system, that have different properties and that interact with each other.[70]*

Jablonka separates these inheritance systems into four groups: genetic, epigenetic, behavioural, and symbolic. Because these systems are continuously interacting with each other this means that the unit of evolutionary selection has to be the organism as a whole. It should be noted that Jablonka is not a full-blown developmental systems theorist because she does not claim that all of the inheritance systems are devoid of information; nevertheless, her analysis is used by developmental systems theorists. I will now outline Jablonka's four groups of inheritance systems.

The genetic inheritance system simply refers to genes: "The gene is a template made up of nucleotides whose sequential organization can be transformed through a complex process of decoding into functional RNA and proteins."[71] The behavioural inheritance system includes inducing-substance transfer, where: "mammal fetuses are able to smell semivolatile liquids transferred to them across the mother's placenta, and later show preference or aversion for food items carrying these smells."[72] They also include both imitation and non-imitative social learning. The symbolic inheritance system also entails social learning and imitation, the difference is that the information here is encoded; this gives symbolic inheritance, "unlimited heredity and huge evolutionary potential."[73]

The epigenetic inheritance systems relate to cellular inheritance between generations, and include the *steady-state system*, *structures inheritance*, and *chromatin-marking*. In contrast to genetic inheritance, epigenetic inheritance is generally patterned and holistic. The *steady-state system* of inheritance is a self-perpetuating system of gene product activity within a cell which is initiated by environmental stimuli. Its presence means that genetically identical cells will be heritably different if their ancestor cells have a different developmental history. These different cellular states, "may be practically unlimited, and cumulative evolutionary change may occur."[74]

Structural inheritance is simply the templating of existing cellular structures to form new structures. Whilst, *chromatin-marks* affect gene expression, this means that genetically identical cells can have heritable differences that can be environmentally induced. Jablonka states that: "The type, the density, and the pattern of marks on a chromosome region affect its potential transcriptional state, and changes in marks can be induced by the changes in the environment."[75]

Griffiths and Gray give an example of how these non-genetic inheritance systems can give rise to fitness differences and adaptation though natural selection. The example involves colonies of the North American fire ant *Solenopsis invicta*. There are no significant genetic differences between the colonies, but some have large, monogynous queens, and others have small, polygynous queens. The way the queen develops depends entirely on the colony in which it is raised – exposure of eggs from a monogynous colony to the pheremonal "culture" of a polygynous colony produces small queens who found further polygynous colonies. This means that, "a mutation in a nongenetic element of the developmental matrix can induce a new self-replicating variant of the system which may differ in fitness from the original."[76]

Utilizing these diverse inheritance systems Griffiths and Gray argue for a developmental systems framework of natural selection that can replace the gene-centered version. In their revised terms: "Developmental system – the interactions and processes that produce a life cycle...Natural selection – the differential reproduction of heritable variants of developmental systems due to relative improvements in their functioning... Adaptation – the product of natural selection...Evolution – change over time in the composition of populations of developmental systems."[77] They argue that this framework widens the scope of Darwinian natural selection to more closely reflect reality:

Not only might expanded forms of inheritance play an important role in the generation of evolutionary novelty they could also significantly alter the dynamics of evolutionary change...expanded inheritance can facilitate transitions from suboptimal to higher peaks, thus creating more effective evolutionary dynamics than would be possible under strict genes-only conceptions. Expanded forms of inheritance may also be the cause of reproductive isolation and hence of speciation.[78]

So, the life-cycle of an organism is constructed through joint-determination by multiple causes. Some of these causes are non-genetic but internal to the organism, these causes are disregarded by the 'modern synthesis' as unimportant; others result from the interaction between the organism and its surroundings, these causes the 'modern synthesis' considers to be a separate and isolated set of processes to those that go on inside the organism. From the holistic perspective of developmental systems theory this separation is an unwarranted and an unhelpful one. No causal factor is in control of the developmental process, as every factor has its effects only in conjunction with feedback effects from the output of other factors. As Oyama states:

the claim is hardly that genetic effects on organisms cannot be identified, but that the genes have their effects by being affected by other factors – by their cellular environments, if you will – and these often include the very processes they influence. The impact of gene products, furthermore, tends to vary with other conditions. Starting an account with genetic transcription, and treating the DNA as an "independent variable" that "initiates" an interesting cascade of events, leads only too easily to obliterating from the causal landscape the events and conditions that brought that transcription about.[79]

Appreciation of the joint reciprocal causation between factors leads us into another claim of developmental systems theory. Whilst the 'modern synthesis' asserts that environmental niches exist independently of organisms, and that random mutations in organisms cause them to 'fit' these niches, developmental systems theory asserts that organisms create their niches. The niche is created both by the organism itself and by its ancestors. Oyama, Griffiths and Gray claim that: "there are no preexisting niches or environmental problems that shape populations from without."[80] They explain this more fully in the following passage:

If evolution is change in developmental systems...it is no longer possible to think of evolution as the shaping of the organism to fit an environmental niche. Rather, the various elements of the developmental systems coevolve. Organisms construct their niches both straight-forwardly by physically transforming their surroundings and, equally importantly, by changing which elements of the external environment are part of the developmental system and thus able to influence the evolutionary process in that lineage.[81]

Another strong advocate of organism-created niches is Richard Lewontin. He claims that: "The concept of an empty ecological niche cannot be made concrete. There is a non-countable infinity of ways in which the physical world can be put together to describe an ecological niche, nearly all of which would seem absurd or arbitrary because we have never seen an organism occupying such a niche."[82] Lewontin asks: "Are there any circumstances in which it can be said that organisms "adapt" to an externally imposed environment rather than "constructing" it by their life activities?[83] He concludes that there are not: "there are no environments without organisms."[84]

So how does an organism construct its environment? Lewontin claims that organisms determine which aspects of the outside world are relevant to them, they actively construct a world around themselves through creating a surrounding atmosphere, and they are also in a constant process of altering their environment. Furthermore, organisms can modulate the statistical properties of external conditions, and they can determine by their biology the actual physical nature of signals from the outside.[85]

To recapitulate, the central claim of developmental systems theory is that the elevation of genes to a prominent role in development is unwarranted. The elevation of genes in the 'modern synthesis' is justified by reference to the Central Dogma of molecular biology which claims that information in DNA and RNA codes for proteins, but that information can never run directly from proteins back to DNA and RNA. The developmental systems theorists reject this view because they assert that information is not something that pre-exists the biological interactions with which it is associated. So, the zygote does not contain DNA sequences that code for proteins. As Peter Godfrey-Smith explains: "The DNA exists in the zygote, but the eventual protein products of that DNA depend on a great array of other components, many of which will only be built up in the course of development."[86]

It is time to move on to 'evolution as path' and ask what we would expect to find in the fossil record if the developmental systems theorists are correct. When the environment plays such a big role in the reconstruction of each life-cycle of an organism it is clear, *ceteris paribus*, that an unchanging environment would not be expected to lead to massive variation or speciation. A moderately changing environment would be expected to lead to increasing variation, and a massive environmental disruption would be expected to lead to relatively rapid speciation.

This means that the expectations of the fossil record would be quite concordant with punctuated equilibrium; we certainly wouldn't expect to see an equal speed of transition between species. We would expect a period of 'equilibrium' wherein organisms are incorporating minor environmental adjustments into their life-cycles, followed by a period of 'punctuation' when big environmental changes occur. Of course, symbiogenesists claim that the punctuated periods are due to symbiogenesis. These two interpretations may actually be concordant, given that the symbioses that lead to symbiogenesis are a product of environment-induced "prototaxis". Direct environment-induced organismic change, which is heritable, is a feature of both of these paradigms, and a feature which is missing from the 'modern synthesis'.

Chapter 4

The state of the Modern Synthesis

The main weakness of modern evolutionary theory is its lack of a fully worked out theory of variation, that is, of candidature for evolution, of the form in which genetic variants are proffered for selection. We have therefore no convincing account of evolutionary progress – of the otherwise inexplicable tendency of organisms to adopt ever more complicated solutions of the problems of remaining alive.[87]

Sir Peter Medawar

As is clear from the above passage, some people have found the attempt of the neo-Darwinian 'modern synthesis' to account for Darwin's missing variation to be far from convincing. I have outlined two contemporary theories which seek to explain the sources of heritable variation without sole reliance on random mutations. Developmental systems theory claims that there are a wide range of non-genetic heritable factors that create evolutionary variation, and that it is the whole life cycle that replicates. As natural selection operates on outcomes, the replicator-vehicle dichotomy becomes meaningless. Symbiogenesis theory, in contrast, claims that the overwhelming majority of variation comes from genome acquisition via symbiogenetic fusions; natural selection works on both the species formed by symbiogenesis and on the trivial variation caused by random mutations, and thereby provides *fine-grained* adaptation.

So, do these two theories pose a serious challenge to the 'modern synthesis'? When it comes to the *path of evolution* it is very hard to pose a serious challenge to the 'modern synthesis' view. Dawkins argues that the fossil record in one place only reveals a 'migration event' rather than an 'evolution event', so we would not expect to find intermediates. This implies either that the intermediates could be found elsewhere, or that we fall back onto the traditional Darwinian claim that the intermediates are missing. So the intermediates are either undiscovered or missing. This is hardly convincing stuff – but it is also irrefutable.

The greater responsiveness of organisms to environmental influences which is hypothesized in developmental systems theory can, perhaps, more adequately explain punctuated equilibrium. This is due to the different cycles of environmental change that occur over geological timescales. Symbiogenesists claim that the punctuations reflect true saltationism – the merging of formerly independent organisms to create a new one. As symbioses originate from behavioural "prototaxis" they would escalate in times of environmental upheaval, and so would also explain the differential speeds inherent to punctuated equilibrium. If evidence continues to accumulate for the symbiotic origins of the eukaryotic cell and for symbiogenetic speciation (as we saw with *Drosophila* and amoebae), then the case for a saltational evolutionary path which encapsulates both branching and fusing phylogenies will become stronger.

The debate over 'evolution as mechanism' whilst appearing to involve three irreconcilable positions, actually appears to be mainly concerned with relative weightings. All three positions accept that random mutations and natural selection play a role in evolution. And all three positions accept that mechanisms other than natural selection and random mutations play a role in evolution. Dawkins admits that:

> *If there are views of the evolution theory that deny slow gradualism, and deny the central role of natural selection, they maybe true in particular cases. But they cannot be the whole truth, for they deny the very heart of the evolution theory, which gives it the power to dissolve astronomi-*

cal improbabilities and explain prodigies of apparent miracle.[88]

It seems that none of the three paradigms in their extreme form is the 'whole truth'. The question is what the relative weighting should be between the following: symbiotic saltationism, expanded inheritance, environmental determination of genetic information, natural selection, and random mutations. Dawkins argues that the overwhelmingly dominant force in the evolutionary mechanism is the natural selection of random genetic mutations, other factors are of minor significance; the environment is important in development, but the genes contain information that has a statistical probability of leading to a given trait. In contrast, developmental systems theorists argue that there is no genetic information without an environment, so the multitude of interacting and heritable environmental factors needs to be given the dominant role in the evolutionary mechanism. Whilst, symbiogenesists reject that the dominant force is natural selection, of either genes, or life-cycles, and claim that the dominant evolutionary mechanism is genome acquisition.

So, do the two challenging theories give a compelling case for taking the dominant role? Margulis and Sagan do make a strong case for increasing the weighting of symbiogenesis in evolution. However, it is possible that the claims that they make are slightly too strong. Firstly, they claim that: "To call the tendency to leave offspring or fail to do so "competition", as biologists frequently do, is misguided."[89] It is hard to see how this statement can fit into any evolutionary mechanism that has a positive role for natural selection; competition is inherent to natural selection.

Secondly, they take a hard line on mutations by claiming that they never lead to speciation – just alteration and refinement: "mutations do not create new species."[90] This claim rests entirely on their definition of a species. David L. Hull claims that, "Species are anything that anyone chooses to make them."[91] This means that an alternative definition of a species – such as 'recognition, mating, and production of fertile offspring' – could lead to the conclusion that an 'alteration' of one species through accumulated mutations actually leads to the creation of a new species. This would be particularly likely in a small isolated island population which would have a high level of inbreeding, and would be isolated from the gene flow of the rest of the species. So, whilst symbiogenesis could be the dominant form of speciation, there could also be speciation without genome acquisition.

The developmental systems theorists also make a strong case for taking the unit of selection to be the organism as a whole, and for abandoning the replicator-vehicle dichotomy. There is surely a large role in generating variation and adaptation for non-genetic heritable factors. Whilst the claim that there is no protein-coding information in DNA that is environment-independent could well turn out to be true. The claim that organisms are jointly determined by multiple causes in a process of self-organization looks set to become increasingly dominant, and it is a view shared by symbiogenesists.

I would suggest that a strong case can be made that the dominant mechanism in evolution is actually a synthesis of the symbiogenesis and developmental systems theories. In the symbiogenetic theory natural selection is utilized in a minor capacity to explain the fine-grained adaptiveness of organisms to their environmental niches; the overwhelming majority of variation, and speciation, comes from symbiogenesis. In developmental systems theory there are no environmental niches that pre-date organisms, as organisms create their own niches.

By combining these two ideas we can postulate an adequate explanation for adaptiveness with no role for positive selection. It is symbiogenesis that generates variation and speciation, and the resulting organisms then create their own niches. We can then explain speciation and adaptiveness without positive natural selection. This view would be the actual fulfilment of the wishes of H. J. Muller who claimed that: "if selection could be somehow dispensed with, so that all variants survived and multiplied, the higher forms would nevertheless have arisen."[92]

It can be concluded that the 'modern synthesis' explanations for the path and the mechanism of evolution are far from complete. Nevertheless, it could be argued that all three of the theories that I have outlined can be synthesized into one overarching theory. Heritable variation would be generated by symbiogenesis, expanded inheritance and random mutations. Whilst, speciation would also occur through symbiogenesis, random mutation accumulation and expanded inheritance.

In this scenario there would be no organisms, or information, without an environment; but given that there is a species-created environment, there is a sense in which the genes that relate to that environment could be described as 'selfish'. This is because a species-created niche creates a high statistical probability that a given gene will be both created and lead to a given trait – this is a 'selfish' attempt to preserve the niche. In concordance with this, we concluded that Dawkins has to fall back onto an equivalence of genotypic with phenotypic traits that is highly contingent on the environmental history of a particular organism – a history which, in the overarching theory, is created by the organism itself. Extended phenotypic effects can be seen as simply a small part of the plethora of organism-environment interactions that constitute the replication of an organism's life-cycle. Rather than one of the three theories being overwhelmingly dominant, it seems that each gives an important perspective on the larger evolutionary picture.

Chapter 5

Do we understand the nature of the universe?

In the previous chapters we have focused solely on biological evolution – the mechanisms of biological evolution, and the paths that biological evolution has taken. In other words, we have focused on life-forms, and how these life-forms have brought forth other life-forms. Now, evolution doesn't just occur at the biological level. If one fully embraces the evolutionary paradigm then one will talk of the evolution of minerals, the evolution of planets and the evolution of solar systems; we can assume that evolution was occurring a long time before the origination of biological evolution.

Compared to biological life-forms, planets and solar systems are vastly larger in size. However, the mechanisms which underpin the formation of planets and solar systems seem to be at work at a vastly *lower* level than the level of biological evolution. It is interactions between the *smallest* constituents of the universe which gives the impetus to the evolution of the universe. The smallest constituents – which on one popular contemporary theory are tiny vibrating 'strings' – give rise to atoms, which, in turn, interact in such a way as to form molecules. As these interactions 'scale up' we end up with biological life-forms and planets. When we talk about biological evolution it is easy to forget that the mechanisms which are operating at this level arise from the mechanisms which are in operation at a much more fundamental level. It is the interactions at the lowest level ('strings'?) – and the mechanisms of evolution which are operating at this level – which underpin evolution at the biological level. In short, the mechanisms of evolution which resulted in the evolution of minerals, planets, solar systems and biological life-forms, are still in existence today. Furthermore, these mechanisms exist today *within* biological life-forms.

Do we understand the nature of these non-biological evolutionary mechanisms? In other words, do we understand the nature of the universe? It seems obvious to me that we don't understand the nature of these mechanisms. We gain access to the universe (our surroundings) via our sensory organs. These organs have been moulded by evolution in a way which causes us to sense the universe in a particular way. It is surely the case that animals with very different sensory organs sense a very different universe.

Not only this, but our sensory organs only give us access to the exterior of things – not their interior. Even if we could see the exterior of the smallest constituents of reality (a 'string'?) we would not know its interior – we would not know if there is anything it is like to be a string; we would not know why a string forms the atom that it does, or why it moves the way that it does.

Clearly, if we don't understand the mechanisms of evolution at this fundamental level, then we cannot fully grasp the forces which underpin biological evolution. Many philosophers have concluded that we do not understand the nature of the universe; that the nature of the universe is mysterious to us. For example, here are some assertions by Immanuel Kant, Peter Unger, Bertrand Russell, and Galen Strawson:

> For every substance, including even a sim-
> ple element of matter, must after all have
> some kind of inner activity as the ground of
> its producing an external effect, and that in
> spite of the fact that I cannot specify in
> what that inner activity consists...Leibniz
> said that this inner ground of all its external
> relations and their changes was a power of
> representation. This thought, which was not
> developed by Leibniz, was greeted with

laughter by later philosophers. They would, however, have been better advised to have first considered the question whether a substance, such as a simple part of matter, would be possible in the complete absence of any inner state.[93]

Except for what little of the physical world we might apprehend in conscious experience, which is available if Materialism should be true, *the physical is mysterious to us.*[94]

we know nothing about the intrinsic quality of physical events except when these are mental events that we directly experience.[95]

I take physicalism to be the view that every real, concrete phenomenon in the universe is ... physical...I will equate 'concrete' with 'spatio-temporally (or at least temporally) located', and I will use 'phenomenon' as a completely general word for any sort of existent. Plainly all mental goings on are concrete phenomena... But how can experiential phenomena be physical phenomena? Many take this claim to be profoundly problematic (this is the 'mind-body problem'). This is usually because they think they know a lot about the nature of the physical. They take the idea that the experiential is physical to be profoundly problematic *given what we know about the nature of the physical.* But they have already made a large and fatal mistake. This is because we have no good reason to think that we know anything about the physical that gives us any reason to find any problem in the idea that experiential phenomena are physical phenomena.[96]

In short, a simple part of matter has an inner state (Kant), the physical is mysterious (Unger), we know nothing about the intrinsic quality of the overwhelming majority of physical events (Russell), and these events could be experiential events (Strawson). Clearly, on this view we are incapable of knowing the fundamental nature of the universe. Therefore, it follows from this that we know very little about the mechanisms of evolution which operate at this level. Given that these mechanisms underpin both the origin of life, and the evolution of biological life-forms from other biological life-forms, it follows that we cannot fully understand the nature of biological evolution.

Of course, we do understand some things about biological evolution. As we have seen in previous chapters, our senses give us access to some parts of the universe and this has enabled us to gain knowledge of some of the kinds of things which are going on within biological life-forms. So, we can be fairly confident that natural selection, developmental systems theory and symbiogenesis all have a role to play in biological evolution.

However, assuming that we do not understand the nature of the universe – that it is 'mysterious' to us – we will lack the knowledge which would enable us to fully understand the nature of biological evolution, and the evolution of the universe in a more general sense. Of course, we can often make accurate predictions (scientific or otherwise) but this doesn't mean that we understand the nature of the phenomena that we are successfully predicting.

The universe, non-biological evolution and biological evolution are all fundamentally mysterious to us, and will remain so in the future. Our understanding of the mechanisms and paths of evolution seems certain to increase; however, we are incapable of demystifying the mysterious universe.

Bibliography

[1] Conway Morris, Simon, *Life's Solution: Inevitable Humans in a Lonely Universe*, (Cambridge: Cambridge University Press, 2005), p. 2.

[2] Jablonka, Eva, and Lamb, Marion J., *Evolution in Four Dimensions*, (London: The MIT Press, 2005), p. 10.

[3] Dawkins, Richard, *The Selfish Gene*, (Oxford: Oxford University Press, 1999), p. 2.

[4] Dawkins, *The Selfish Gene*, p. 36.

[5] Dawkins, *The Selfish Gene*, p. 33.

[6] Dawkins, *The Selfish Gene*, p. 87.

[7] Dawkins, *The Selfish Gene*, p. 44.

[8] Dawkins, *The Selfish Gene*, p. 205.

[9] Dawkins, *The Selfish Gene*, p. 204.

[10] Dawkins, *The Selfish Gene*, p. 274.

[11] Dawkins, Richard, *The Extended Phenotype*, (Oxford: Oxford University Press, 1999), p. 38.

[12] Dawkins, *The Extended Phenotype*, p. 28.

[13] Dawkins, *The Extended Phenotype*, p. 195.

[14] Dawkins, *The Extended Phenotype*, p. 210.

[15] Dawkins, *The Extended Phenotype*, p. 218.

[16] Dawkins, *The Extended Phenotype*, p. 231.

[17] Dawkins, *The Extended Phenotype*, p. 227.

[18] Dawkins, *The Extended Phenotype*, p. 233.

[19] Dawkins, Richard, *The Blind Watchmaker*, (London: Penguin Books, 2006), p. 43.

[20] Dawkins, *The Blind Watchmaker*, p. 234.

[21] Dawkins, *The Blind Watchmaker*, p. 233.

[22] Dawkins, *The Blind Watchmaker*, p. 234.

[23] Dawkins, *The Blind Watchmaker*, p. 193.

[24] Dawkins, *The Blind Watchmaker*, p. 193.

25 Dawkins, *The Blind Watchmaker*, p. 230.
26 Dawkins, *The Blind Watchmaker*, p. 241.
27 Dawkins, *The Blind Watchmaker*, p. 241.
28 Dawkins, *The Blind Watchmaker*, p. 239.
29 Dawkins, *The Blind Watchmaker*, pp. 237-9.
30 Margulis, Lynn, and Sagan, Dorion, *Acquiring Genomes: A theory of the origins of species*, (New York: Basic Books, 2002), p. 12.
31 Margulis and Sagan, *Acquiring Genomes*, p. 12.
32 Dawkins, *The Selfish Gene*, p. 172.
33 Dawkins, *The Selfish Gene*, pp. 181-2.
34 Dawkins, *The Selfish Gene*, p. 182.
35 Dawkins, *The Extended Phenotype*, pp. 222-3.
36 Margulis and Sagan, *Acquiring Genomes*, p. 11.
37 Margulis, Lynn, *The Symbiotic Planet: A new look at evolution,* (Guernsey: Phoenix, 2001), p. 33.
38 Margulis and Sagan, *Acquiring Genomes*, pp. 11-12.
39 Margulis and Sagan, *Acquiring Genomes*, p. 12.
40 Margulis and Sagan, *Acquiring Genomes*, p. 29.
41 Dawkins, *The Selfish Gene*, p. 19.
42 Dawkins, *The Selfish Gene*, p. 46.
43 Margulis, *The Symbiotic Planet*, pp. 91-2.
44 Margulis and Sagan, *Acquiring Genomes*, pp. 55-56.
45 Margulis and Sagan, *Acquiring Genomes*, p. 56.
46 Margulis and Sagan, *Acquiring Genomes*, p. 72.
47 Margulis and Sagan, *Acquiring Genomes*, p. 89.
48 Margulis and Sagan, *Acquiring Genomes*, p. 94.
49 Margulis and Sagan, *Acquiring Genomes*, p. 99.
50 Margulis and Sagan, *Acquiring Genomes*, p. 102.
51 Margulis and Sagan, *Acquiring Genomes*, p. 102.
52 Margulis and Sagan, *Acquiring Genomes*, p. 157.
53 Margulis and Sagan, *Acquiring Genomes*, p. 165.
54 Margulis and Sagan, *Acquiring Genomes*, p. 166.

55 Margulis and Sagan, *Acquiring Genomes*, p. 169.

56 Margulis and Sagan, *Acquiring Genomes*, p. 169.

57 Margulis and Sagan, *Acquiring Genomes*, p. 180.

58 Margulis and Sagan, *Acquiring Genomes*, p. 192.

59 Margulis, *The Symbiotic Planet*, p. 9.

60 Margulis, Lynn, and Sagan, Dorion, *Microcosmos: Four billion years of microbial evolution*, (London: University of California Press, 1997) pp. 121-2.

61 Margulis and Sagan, *Acquiring Genomes*, p. 15.

62 Margulis and Sagan, *Acquiring Genomes*, p. 172.

63 Margulis and Sagan, *Acquiring Genomes*, p. 96.

64 Margulis and Sagan, *Acquiring Genomes*, p. 83.

65 Margulis and Sagan, *Acquiring Genomes*, p. 201.

66 Oyama, Susan, Griffiths, Paul E., and Gray, Russell D. (Ed.), *Cycles of Contingency*, (London: The MIT Press, 2001) p. 1.

67 Oyama, Griffiths, and Gray, *Cycles of Contingency*, p. 1.

68 Oyama, Susan, "Terms in Tension: What Do You Do When All the Good Words Are Taken", in Oyama, Griffiths, and Gray, *Cycles of Contingency*, p. 179.

69 Weber, Bruce H., and Depew, David J., "Developmental Systems, Darwinian Evolution, and the Unity of Science", in Oyama, Griffiths, and Gray, *Cycles of Contingency*, p. 240.

70 Jablonka, Eva, "The Systems of Inheritance", in Oyama, Griffiths, and Gray, *Cycles of Contingency*, p. 99.

71 Jablonka, Eva, "The Systems of Inheritance", p. 100.

72 Jablonka, Eva, "The Systems of Inheritance", p. 110.

73 Jablonka, Eva, "The Systems of Inheritance", p. 112.

74 Jablonka, Eva, "The Systems of Inheritance", p. 105.

75 Jablonka, Eva, "The Systems of Inheritance", p. 106.

76 Griffiths, Paul E., and Gray, Russell D., "Darwinism and Developmental Systems", in Oyama, Griffiths, and Gray, *Cycles of Contingency*, p. 199.

77 Griffiths and Gray, "Darwinism and Developmental Systems", p. 214.

78 Griffiths and Gray, "Darwinism and Developmental Systems", p. 200.

79 Oyama, "Terms in Tension: What Do You Do When All the Good Words Are Taken", p. 182.

80 Oyama, Susan, Griffiths, Paul E., and Gray, Russell D. (Ed.), *Cycles of Contingency*, p. 6.

81 Oyama, Susan, Griffiths, Paul E., and Gray, Russell D. (Ed.), *Cycles of Contingency*, p. 6.

82 Lewontin, Richard, *The Triple Helix*, (London: Harvard University Press, 2000) p. 49.

83 Lewontin, *The Triple Helix*, p. 66.

84 Lewontin, *The Triple Helix*, p. 67.

85 Lewontin, *The Triple Helix*, pp. 51-63.

86 Godfrey-Smith, Peter, "On the Status and Explanatory Structure of Developmental Systems Theory", in Oyama, Griffiths, and Gray, *Cycles of Contingency*, p. 293.

87 Medawar, Sir Peter, cited in Dawkins, *The Extended Phenotype*, p. 165.

88 Dawkins, *The Extended Phenotype*, p. 318.

89 Margulis and Sagan, *Acquiring Genomes*, p. 17.

90 Margulis and Sagan, *Acquiring Genomes*, p. 72.

91 Hull, David L., "Introduction to Part IV: Species" in Hull, David L., and Ruse, Michael, *The Philosophy of Biology*, (Oxford: Oxford University Press, 1998), p.297.

92 Muller, H. J., cited in Ruse, Michael (Ed.), *Philosophy of Biology*, (New York: Macmillan Publishing Company, 1989), p. 90.

93 Kant, Immanuel, "Dreams of a Spirit-Seer", in *Theoretical Philosophy 1755-1770*, ed. David Walford, (New York: Cambridge University Press, 1992), p. 315.

94 Unger, Peter, *All the Power in the World*, Oxford, Oxford University Press, 2006, p. 5.
95 Russell, Bertrand, *Portraits from Memory*, Spokesman, 1956, p. 153.
96 Strawson, Galen, *Consciousness and its place in nature,* Imprint Academic, Exeter, 2006, pp. 3-4.

www.ingramcontent.com/pod-product-compliance
Lightning Source LLC
Chambersburg PA
CBHW060644210326
41520CB00010B/1735